CONTENTS

All the words that appear in **bold** type are explained in Words we use on page 30.

BITTER COLD

Cold **weather** makes water freeze. **Ice, frost** and snow are forms of frozen water. They glitter and sparkle and they are all slippery. Children like to play sliding games on ice.

After playing outside, it feels good to go back inside a warm house. People need warmth to stay alive. We must protect our bodies against the cold.

This man is carving a statue in ice. His saw has teeth which grip the slippery surface of the ice.

These children in Japan are waiting to slide on the ice. Sometimes we can see through ice, like glass. It is transparent. Sometimes ice is white and full of tiny bubbles of air. Ice is hard and usually has a smooth, wet surface.

ROTHERHAM LIBRARY & INFORMATION SERVICES

This book must be returned by the date specified at the time of issue
as the DATE DUE FOR RETURN.
The loan may be extended (personally, by post or telephone) for a
further period if the book is not required by another reader, by quoting
the above number / author / title.

LIS7a

ICE AND PEOPLE

NIKKI BUNDEY

A ZOË BOOK

A ZOË BOOK

© 2000 Zoë Books Limited

Devised and produced by
Zoë Books Limited
15 Worthy Lane
Winchester
Hampshire SO23 7AB
England

First published in Great Britain in 2000 by
Zoë Books Limited
15 Worthy Lane
Winchester
Hampshire SO23 7AB

A record of the CIP data is available from the British Library.

ISBN 1 86173 027 6

Printed in Italy by Grafedit SpA
Editor: Kath Davies
Design: Sterling Associates
Illustrations: Artistic License/Genny Haines,
   Tracy Fennell, Janie Pirie
Production: Grahame Griffiths

Photographic acknowledgments
The publishers wish to acknowledge, with thanks, the following photographic sources:

5 Mary Evans Picture Library/Arthur Rackham Collection; Liba Taylor 10t / Liz McLeod 12t / John Downman 25t / Andrey Zvoznikov 26t / Hutchison Picture Library; Ken Graham - title page / Material World 10b / Simon Shepheard 14t / Charles Worthington 19t / Impact Photos; A.N.T. 6 / B & C Alexander 11, 20b, 28 / Rich Kirchner 17b / Kevin Schafer 21 / N.A.Callow 23t / NHPA; Andre Maslennikov - cover (background) / B & C Alexander - cover (inset) left, 13, 22, 23b / Hartmut Schwarzbach 26b / Still Pictures; cover (inset) right, 7, 12b, 17t, 18, 20t, 24, 27, 29 / The Stock Market; M Watson 4t & b / Viesti Collection 8 / H Rogers 9, 16 / B Devine 14b / M Barlow 15 / G Gunnarsson 19b / J Moscrop 25b /TRIP.

The publishers have made every effort to trace the copyright holders, but if they have inadvertently overlooked any, they will be pleased to make the necessary arrangement at the first opportunity.

Jack Frost was a mean enemy and could hurt people in the stories told about him. Today we understand more about the scientific reasons for cold **temperatures**. But people still die when the weather turns very cold.

In the days before homes had **central heating**, people were often very cold. They told children stories in which the ice and frost were like real people. Father Frost was a Russian Santa Claus. Jack Frost was wicked and cruel.

Jack Frost's fingernails were made of **icicles**. He pinched everyone's fingers and toes. When we are very cold, this is just how our fingers and toes feel.

DOWN TO ZERO

Cold affects our health and how we live. It affects our **environment**, the plants we grow and the food we eat.

Sometimes we need to measure the temperature of air, water or the human body. To do this, we use instruments called **thermometers**. They record temperature in units called **degrees** (°). There are different **scales** of temperature, such as Celsius (C) or Fahrenheit (F). Water freezes at 0°C or 32°F.

The Arctic and the Antarctic are the coldest places on Earth. The cold wind makes them feel even colder. It lowers the temperature of the human body, too. This effect is called wind chill.

Air warms up on the Earth's surface. It rises and then quickly cools. The higher the land, the colder the air gets. Mount Everest is the world's highest mountain. There, people climb in extreme cold.

Why do we get cold weather? The Sun's rays fall on the Earth at a different slant, or angle, in different places. People living near the **Equator** enjoy warm sunshine, but **polar** explorers live in bitter cold.

The Earth tilts as it travels round the Sun. As a region on Earth tips away from the Sun, it has a winter **season**. As a region tips towards the Sun, it has summer.

See for Yourself

- Use a thermometer to record the air temperature at the same time of day, every day for one week.

- Measure the air temperature at the same time of night for one week.
- What is the average difference between the two daily temperatures?
- Why do temperatures fall at night?

ICE, WATER AND US

People, like animals and plants, need water to stay alive. Water covers two thirds of our **planet**, Earth.

The water in rivers, lakes and the salty oceans is a **liquid**. The great ice caps which cover the **North Pole** and the **South Pole** are **solid** water. As a **gas** called **water vapour**, water is in the air we breathe.

Dripping water on Mount Rainier, in Washington State, USA, has frozen solid to form long icicles. When the temperature rises, the ice will turn, or melt, back into liquid water.

When the Sun heats the oceans, the water **evaporates** and turns into water vapour. Like other gases, water vapour rises when it is warm. High in the **atmosphere** it cools and **condenses**. It turns back into liquid **droplets** or **ice crystals**. These may fall from the clouds as rain or snow. Sometimes they form frost at ground level, or water drops called dew.

Ice, snow and frost can make us cold, but they are part of the **water cycle** on this planet.

Liquid water flows and drips. It is fluid. Solid ice is hard and cannot move easily. Water vapour, a gas, can move freely, filling up any space.

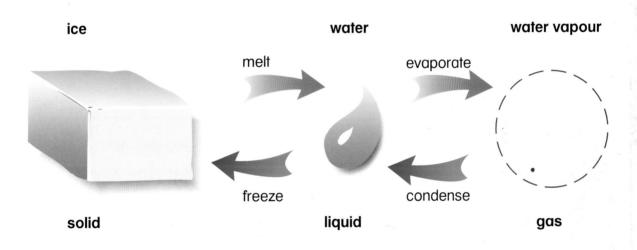

ice   **water**   **water vapour**

melt   evaporate

freeze   condense

solid   **liquid**   **gas**

When water freezes, tiny ice crystals build up and stick together. Most of the world's **fresh water** is frozen water found in the polar ice caps.

STAYING ALIVE

Our bodies are designed to survive in the cold. When we eat our bodies turn food into **energy,** which warms us. Energy makes our muscles work. Moving around also raises the body's temperature.

The human body works well only if its temperature stays around 37°C. A group of **nerve cells** in the brain works like a central heating **thermostat.** The cells control the flow of blood around our bodies to keep us warm or cool.

In cold weather goose bumps raise the small hairs on our bodies. They trap a layer of warm air to keep us warm. In an icy Russian winter, people need more layers of warmth, so they wear more clothes and hats.

The human body can stand some changes in skin temperature. This girl has been swimming outdoors in Iceland. The pool is warm, but there is snow all around it.

Narrow eyes protect people from the blinding dazzle of white ice and from freezing winds. These Inuit children are from Baffin Island in Canada.

If people have no food, if they cannot move around or if they fall into cold water, their body temperatures fall. At danger level, this state is called **hypothermia**. Babies and old people are especially at risk.

To prevent heat loss, the brain's thermostat may reduce the supply of blood to the skin's surface. But then our flesh may freeze. Damage caused to skin and flesh by freezing is called frostbite.

See for Yourself

- Learn how to take a friend's temperature using a thermometer. **Ask an adult to help you do this.**
- Take the temperature when your friend is in good health, on a cold day.
- Do the same on a hot day.
- Is there much difference between the two readings?
- Do you think the thermostat in your friend's brain is doing its job?

FIGHT AGAINST COLD

We can protect our bodies from the cold and make, or generate, our own warmth. We may drink hot tea or cocoa or eat food to warm up. We may run or do hard, physical work. We may jump up and down or flap our arms to help the blood to move around, or circulate, through our bodies.

We can raise the air temperature with a fire, a stove or a radiator. We can prevent heat escaping from our bodies by **insulating** them with layers of clothing.

These Bosnian boys are on their way to play ice hockey. They will use their muscles and will soon warm up on this cold day.

Climbers wear special clothing to protect their bodies in icy conditions. These clothes can insulate the body and reduce wind chill. **Waterproof** layers keep the climbers dry.

A strong wind blowing over ice can cause frostbite. We can shield our bodies from the chilling wind by building barriers. People make windbreaks or shelters from cloth, wood or stone. Even a wall of ice or snow will help keep out the cold.

This Cree woman lives in Quebec province, in Canada. The picture shows four ways in which she is fighting the cold. Firstly, she wears a warm coat and gloves. Secondly, she is keeping moving, chopping logs. Thirdly, she can burn the logs to make a fire. Fourthly, she has raised a shelter in which to live.

BUILDING FOR WINTER

Buildings are insulated to protect us against the cold. The walls have two layers with a gap in between. The gap traps a layer of warm air around the house. Window glass may also have two layers, or double glazing. Builders line attics with fibre to stop warmth escaping through the roof. They also wrap or lag water pipes to stop them from freezing.

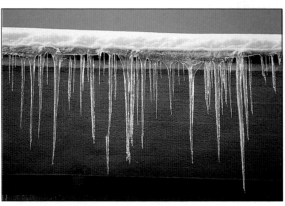

Icicles drip from a roof in Zakopane, Poland. In countries which have cold winters, roofs must be strong enough to bear the weight of heavy ice and snow. Gutters and drains must hold a rush of water during the spring thaw.

As ice **expands**, it can crack a frozen water pipe. When the ice melts, water bursts through the pipe and causes a flood. Here the water from a burst pipe has frozen again, making huge icicles.

This timber house is near Lake Baikal in Siberia, where winters are harsh. The floor is a deck, raised above the frozen soil. Concrete blocks anchor the building to the ground.

Ice and frost can damage building materials. Clay, brick and stone all soak up water. If the water freezes, it expands and takes up more space, causing cracks and crumbling.

In the same way, freezing and thawing of water and ice break up the soil. In the Arctic **tundra**, a deep layer of soil stays frozen as **permafrost** all year round. Only the layer of soil near the surface thaws in summer. Landslips are common. If the **foundations** of houses slip, or **subside**, the building may fall down.

TRAVELLING AROUND

Slippery pavements and roads make travel difficult when the weather is cold. People may fall and break arms or legs. Cars may skid and crash.

The rubber soles of boots and the tyres of cars have ridges called treads. They grip the surface with a rubbing force called **friction**. When there is enough grip for a vehicle to move forward, it is called traction.

Trucks drop grit or salt on roads in winter. Grit makes icy surfaces rougher. It increases the friction between ice and tyres. Salt lowers the temperature at which water can freeze.

In polar regions, seas freeze over for months each year. Ships called ice-breakers smash channels through the ice. Their powerful engines and strong fronts, or bows, push away the great weight of ice.

This aircraft has skis, so that it can make a controlled slide over the ice when it lands. Wheels are not much use on the Antarctic ice cap.

Ice is dangerous. It freezes **points** which re-route trains. If ice forms on a ship in cold seas, the ship may become **top-heavy**, overturn and sink. Aircraft may also ice up in the cold at high **altitude**.

Vehicles, engines and tracks must be ice-free if they are to work in winter. People use chemical mixtures such as **antifreeze** to stop liquids from freezing. Electric heating thaws frosty windscreens and frozen railway tracks.

PLANTS IN THE COLD

Frost and ice can help farmers. In the fields, clods of **waterlogged** soil break up and crumble when they freeze. It is easier for farmers to sow seeds in the crumbled soil when spring arrives.

Frost can also ruin crops. Farmers fear that frost may arrive earlier or later than usual. If it does, it can freeze and destroy tender shoots, buds and leaves.

Freezing conditions can make the fields rock hard. Plants cannot take up frozen water through their roots.

18

These crops are growing inside a big plastic tube called a polytunnel. It protects the plants from frost damage.

Farmers sometimes insulate the ground and the young shoots from frost. They may spread a layer of straw over them to protect, or **mulch**, them. Farmers may cover the fields in plastic sheeting.

Some plants are protected under glass, inside a greenhouse or a cold frame. Sometimes fruit growers light slow bonfires or gas burners in their orchards at night. This heat prevents frost from forming.

Iceland is near the Arctic, so farmers there grow flowers in heated greenhouses. The heating power comes from deep below the Earth's surface. It is called geothermal power, and comes from volcanic activity.

IN THE FRIDGE

Tiny living things called bacteria can make food rot. Bacteria cannot spread easily when temperatures are below **freezing point**.

Long ago, people discovered that they could use cold temperatures to **preserve** foods. The ancient peoples of the Andes mountains learned how to **freeze-dry** potatoes. They left potatoes out in the sunshine and severe frost. Before **refrigerators** were invented, people used blocks of ice to keep food cold and fresh.

Fish will keep fresh if they are packed in ice. On big trawlers, the fish are deep-frozen on board ship.

The Arctic ice acts as a giant freezer. The Inuit people of Greenland use ice to preserve the animals they hunt.

Ice helps to keep these salmon free from bacteria. The bacteria could make people ill.

Scientists realised that some liquids cause cooling when they evaporate. These liquids are used to make refrigerators and freezers work.

Today, there are refrigerators and freezers in homes around the world. They keep food cold to preserve it. Hospitals use refrigerators to store blood and medicine. Freezing is also used in factories.

See for Yourself

- Look through your shopping trolley. Read the labels on the food you have chosen.
- Which items will you put in the fridge? How long can they stay there safely?
- Which items should go into the freezer? How long can they be stored there safely?
- Which items do not need cold storage? How are they preserved?
- Are there any items in the fridge or freezer which the cold might spoil?

LIFE IN ICY LANDS

Scientists have set up bases in Antarctica, but no one has ever made a home in that icy land. Conditions are just too harsh.

In other places, humans have learned to live with ice and frost. In Russian Siberia, frozen rivers can support heavy trucks. These rivers are used as roads. People have learned to fish through holes in frozen lakes. Some Arctic peoples live by hunting, others by herding reindeer.

The Inuit people of the Arctic invented a light, skin-covered canoe called a kayak. It is easy to carry, and if ice damages it during a hunting trip, it is easy to repair. The kayak has a slender, **streamlined** shape. It slips through the water with little **resistance**.

The people of Tibet live in the Himalayan mountains. They keep yaks. These shaggy-coated oxen can put up with extreme cold. They carry heavy loads through icy mountain passes. Yaks provide milk which is used to make butter.

Many people who herd animals live on mountainsides. They often spend the harsh winters in the shelter of the valleys. Here there are enough trees to provide fuel for their fires. When the ice melts in spring, they take their herds to graze the high slopes above the **tree line**.

These boys are taking ice back to their village in Greenland, where it will be melted and used for cooking and cleaning.

ICE SPORTS

Many people welcome icy weather. They take part in ice sports such as ice-skating. The first skates were made of wood or bone, but today they usually have steel blades.

People still skate on lakes and rivers, but mostly they go to ice rinks. Their blades slide easily over the smooth ice. There is not much friction because a thin film of water lies on the surface of the ice.

These women are playing ice hockey. It is a tough sport played on skates. They wear padding and gloves. These cushion the **impact** if they fall or bump into one another.

The weight of the skater's body gives it forward thrust, or **impetus**. The skaters must balance, or the downward pull of **gravity** will send them sprawling across the ice.

Figure skaters twist, turn and dance on the ice. They need extra friction to control their movements. There are small notches at the front of their blades, which they can dig into the ice.

Speed skaters hunch down and wear tight-fitting clothes so that they slip through the air easily. They swing their arms to thrust their bodies forwards. They can travel at 50 kilometres per hour.

All sorts of sports take place on ice. These sleds, or ice yachts, are on the Songhua river in China. They slide over the surface at very high speeds. The force of the wind in the sails pushes the yachts forward.

FROST ON THE WAY?

Most of us like to know when cold weather is on the way. Gardeners and farmers need to know if there will be heavy frost. Icy conditions threaten the lives of drivers, mountain climbers, sailors and airline pilots.

People sometimes use old sayings to **forecast** the weather. 'As the days get longer, the cold gets stronger', is one true saying. It means that after a long winter, the soil stays chilled through early spring.

The scientific study of weather is called **meteorology**. Weather stations report conditions around the world. In space, **satellites** keep track of weather systems as they sweep through the Earth's atmosphere.

Some people's lives depend on the weather. They learn to read natural signs, such as clouds and winds. This Siberian boy is looking after a young reindeer in early spring. He needs to know whether there will be a freeze or a thaw.

These backpackers are in the Himalayan Mountains. They need a detailed weather forecast in order to stay safe. They cannot rely on guesswork or old sayings.

Pilots never take off without checking a weather report. Instruments in the cockpit show the temperature outside the aircraft. They also show the altitude.

The force of air pressing down on the Earth is called **air pressure**. Areas of low air pressure are called depressions. They bring clouds, so we expect rain or snow. High pressure areas are called anticyclones. These bring clear weather, without clouds. With no clouds to trap the Earth's heat, the air turns cold. Frost may form near the ground.

See for Yourself

- Note the details of the television weather forecast every day for one month in winter.
- Keep a record of your local weather conditions each day.
- How often was the forecast correct? How often was it wrong?

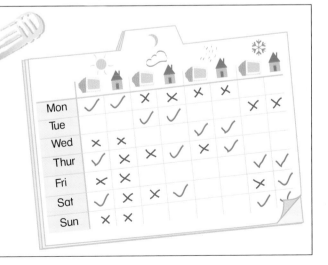

PAST, PRESENT, FUTURE

The amount of frozen water on Earth slowly changes over long periods. In past **Ice Ages**, the ice cap stretched far beyond the polar regions. During warmer periods, the ice cap melted. More water filled the oceans and sea levels rose.

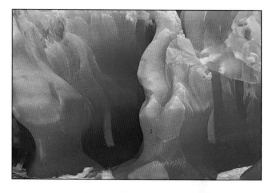

Climate change is important to human life. It affects the plants we grow and eat, and the animals we hunt and herd. It may reduce or increase the areas where we can live.

Sometimes large chunks of ice break away from polar **glaciers** or **ice-shelves** and float out to sea. These chunks are called **icebergs**. When temperatures rise, more icebergs break free.

This print shows a 'frost fair' in London, more than 300 years ago. It was held on the frozen River Thames. At that time it was much colder in northern Europe than it is today.

The world's climate is changing faster than ever. Chemicals have **polluted** the atmosphere and heated up our planet. This process is called **global warming**.

Global warming could be disastrous. There might be more stormy weather and massive flooding. Climate change could lead to poorer harvests of wheat and rice, and create food shortages. Our future may be decided by the effects of global warming.

Scientists check on global warming on the coasts of Antarctica. Many people believe that pollution is causing climate change.

WORDS WE USE

air pressure	The force of the atmosphere pressing down on the Earth's surface.
altitude	Height above sea level.
antifreeze	Chemicals mixed with liquids to reduce their freezing points.
atmosphere	The layer of gases surrounding a planet.
central heating	A system of radiators in a building warmed by a central boiler.
climate	The pattern of weather in one place over a long period.
condense	To turn from gas into liquid.
degree	A unit on a scale, used in measuring temperatures.
droplet	A tiny drop. Droplets combine to form a raindrop or snowflake.
energy	The power to carry out an action.
environment	The world around us, our surroundings.
Equator	An imaginary line around the middle of the Earth.
evaporate	To turn from liquid into gas.
expand	To take up more space, to grow bigger.
forecast	To try to work out, or predict, what the weather will be.
foundations	The parts of a house which support it, usually below ground.
freeze-dry	To preserve food by drying and freezing.
freezing point	The temperature at which water turns to ice (0°C).
fresh water	Water with no salt content.
friction	The force which slows an object as it rubs against another.
frost	A covering of ice needles or an air temperature below 0°C.
gas	An airy substance which fills any space in which it is contained.
glacier	A river of ice, made from hard-packed snow.
global warming	The warming of the Earth.
gravity	The force which pulls objects towards the Earth's surface.
hypothermia	The cooling of the body to very low temperatures.
ice	Water frozen into a solid state.
Ice Age	A time of global cooling, when ice spreads far from polar regions.
iceberg	A large slab of ice which floats through the sea.
ice crystal	A tiny particle of ice which forms in a regular pattern.
ice-shelf	A large ledge of ice extending from a coastline.

icicle	A large needle of ice formed by freezing drops of water.
impact	The force with which one object strikes another.
impetus	The forward thrust of a moving object.
insulating	Using a barrier to prevent a loss of heat, sound or electricity.
liquid	A fluid substance such as water.
meteorology	The scientific study of weather conditions.
mulch	A layer of leaves or straw which protects plants.
nerve cells	Fibres in the body which carry messages to and from the brain.
North Pole	The most northerly point on Earth.
permafrost	A deep layer of frozen soil which never melts.
planet	One of the worlds travelling around the Sun, such as Earth.
points	The places where rail tracks change trains' direction.
polar	To do with regions surrounding the North and South Poles.
polluted	Poisoned or made impure.
preserve	To keep fresh.
refrigerator	A machine used to store food and keep things cold.
resistance	Friction which slows an object as it travels.
satellite	A spacecraft or other object which circles a planet.
scale	A series of graded units.
season	Climate variation caused as the Earth tilts and travels round the Sun. Spring, summer, autumn and winter are seasons.
solid	A hard substance with three dimensions (length, height, width).
South Pole	The most southerly point on Earth.
streamlined	Designed to slip easily through air or water.
subside	To sink to a lower level.
temperature	Warmth or coldness, measured in degrees.
thermometer	A device for measuring the temperature of air, water or our bodies.
thermostat	A device for controlling the temperature of a machine or a heater.
top-heavy	Unbalanced, heavier at the top than the bottom.
tree line	The upper limit of tree growth on a mountainside.
tundra	Treeless regions bordering the polar ice.
water cycle	The ongoing process in which rain falls, evaporates, rises and condenses.
waterlogged	Soaked and unable to hold any more liquid.
waterproof	Keeping out or repelling water.
water vapour	A gas created when water evaporates.
weather	Atmospheric conditions such as heat, cold, sun, rain and snow.

INDEX